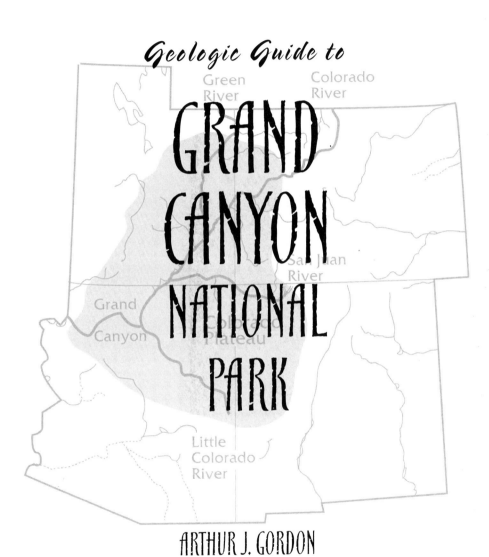

Geologic Guide to

GRAND CANYON NATIONAL PARK

ARTHUR J. GORDON

KENDALL/HUNT PUBLISHING COMPANY
4050 Westmark Drive   Dubuque, Iowa 52002

Front cover image courtesy of Corel.

Inside front cover image adapted from Collier, *An Introduction to Grand Canyon Geology*

Copyright © 2000 by Kendall/Hunt Publishing Company

ISBN 0-7872-6510-1

Library of Congress Catalog Card Number: 00-102125

All rights reserved. No part of this publication may be reproduced, stored in a retrieval system, or transmitted in any form or by any means, electronic, mechanical, photocopying, recording, or otherwise, without the prior written permission of the copyright owner.

Printed in the United States of America
10 9 8 7 6 5 4 3 2 1

CONTENTS

The Geologic Time Scale iv

Preface v

1. Introduction 1
2. Stratigraphy 9
3. Metamorphic and Igneous Rocks 27
4. Structural Geology 35
5. Geomorphology 43
6. Groundwater and Springs 53
7. Geologic History 61

Glossary 71

References 75

THE GEOLOGIC TIME SCALE

ERAS	PERIODS		EPOCHS	LENGTH IN MILLIONS OF YEARS	AGE ESTIMATES OF BOUNDARIES IN MILLIONS OF YEARS
Cenozoic	Quaternary		Holocene	.01	.01
			Pleistocene	1.6	1.6
	Tertiary	Neogene (subperiod)	Pliocene	3.7	5.3
			Miocene	18.4	23.7
		Paleogene (subperiod)	Oligocene	12.9	36.6
			Eocene	21.2	57.8
			Paleocene	8.6	66.4
Mesozoic	Cretaceous			78	144
	Jurassic			64	208
	Triassic			37	245
Paleozoic	Permian			41	286
	Pennsylvanian			34	320
	Mississippian			40	360
	Devonian			48	408
	Silurian			30	438
	Ordovician			67	505
	Cambrian			65	570
	Precambrian time*				
Proterozoic Eon					2500
Archean Eon					
	Oldest known rocks in the United States				3600

*A general consensus has not been arrived at regarding divisions of Precambrian time in North America and elsewhere in the world. The terms given here are considered informal; that is, time terms without specific rank. An Eon is the longest division of geologic time.

PREFACE

As a geology instructor at the community college level, I take great pleasure in teaching introductory geology courses. I remember taking earth science in ninth grade, and for the first time, enjoyed a class so much that I looked forward to it each day. The material was interesting, and the subject matter made sense. I began to understand what was beneath my feet, and I could visualize three-dimensional relationships.

It was with this enthusiasm that I undertook the writing of this book. Most people visiting the Grand Canyon are not geologists, but tourists appreciating one of the great wonders of nature. This book has been written for them and for others who plan to visit the area. I have attempted to explain basic geologic principles, and the geology of the Grand Canyon in terms that people of different ages and interests can understand.

FEATURES OF THIS BOOK

- Geologic terms highlighted in bold are defined in the glossary.
- Figures and photographs help readers visualize geologic principles and three-dimensional relationships.

ACKNOWLEDGMENTS

I would like to acknowledge the thoughtful review of this book by geologists David E. Wahl, Jr., Ph.D., and Esther Tuttle, and the fulfilling support of my wife, Sara T. Gordon, who encourages all of my endeavors.

INTRODUCTION

The Grand Canyon is the second most-visited national park in the United States. People come from around the world to visit the Grand Canyon, be it at the South Rim as a quick side visit on the way to Las Vegas, or a two-week raft trip down the Colorado River. The fascination with the Grand Canyon is never ending. This book attempts to help readers understand the Grand Canyon, the rocks exposed on its walls, and the history of its development.

Before getting into the geology of the Grand Canyon, it is important to understand the regional geologic setting. As big as the Grand Canyon is, it is but a minor feature, geologically speaking, of a larger area known as the Colorado Plateau.

Throughout the United States, and worldwide for that matter, there are notable differences among landscapes. Each region has its own set of characteristic landforms that have developed as a result of the underlying geology. Similarity in landforms is thus explained by similar geologic conditions. The continental United States can be divided

2 Introduction

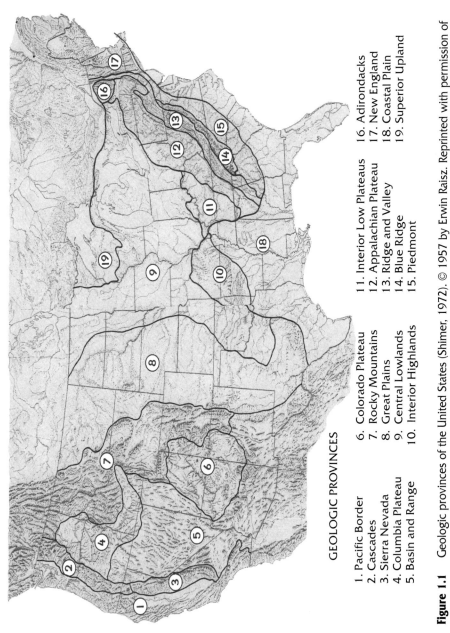

Figure 1.1 Geologic provinces of the United States (Shimer, 1972). © 1957 by Erwin Raisz. Reprinted with permission of RAISZ LANDFORM MAPS, Brookline, Massachusetts. 800-277-0047

GEOLOGIC PROVINCES

1. Pacific Border
2. Cascades
3. Sierra Nevada
4. Columbia Plateau
5. Basin and Range
6. Colorado Plateau
7. Rocky Mountains
8. Great Plains
9. Central Lowlands
10. Interior Highlands
11. Interior Low Plateaus
12. Appalachian Plateau
13. Ridge and Valley
14. Blue Ridge
15. Piedmont
16. Adirondacks
17. New England
18. Coastal Plain
19. Superior Upland

Introduction **3**

into 19 geologic provinces (Figure 1.1). A **geologic province** is a large region characterized by similar geologic history and development. Some boundaries between provinces are obvious and abrupt, such as where the Rocky Mountains tower over the Great Plains. Some boundaries may interfinger with one another, and others may be gradational and not easily recognized.

The Grand Canyon is located within the Colorado Plateau geologic province (Figure 1.2), which covers approximately 150,000

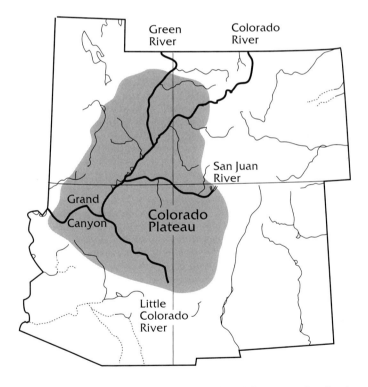

Figure 1.2 Map showing the Colorado Plateau, Grand Canyon, and major rivers (adapted from Collier, AN INTRODUCTION TO GRAND CANYON GEOLOGY).

square miles in the "four corners" states of Arizona, Utah, Colorado, and New Mexico. A **plateau** is an elevated area of relatively flat land that contains somewhat abrupt descents along much of its perimeter. The Colorado Plateau meets this definition, as its surface is generally over one mile above sea level except in narrow, deeply eroded canyons, such as the Grand Canyon. The Rocky Mountains lie to the north and east of the Colorado Plateau, which descends to the Basin and Range province to the southeast, south, and west.

Horizontal layers of thick sedimentary rocks that have been uplifted dominate the Colorado Plateau. **Sedimentary rocks** are composed of weathered and eroded remnants of pre-existing rock that have been redeposited in a horizontal manner, and are thus layered, or **stratified**. This stratification of differently colored sedimentary rocks gives the Grand Canyon its unique banded appearance. Fossils found in many sedimentary layers are useful for dating these rocks. The Grand Canyon sedimentary rocks, along with their fossils, are described in Chapter 2.

Although the elevation of the surface of the Colorado Plateau varies somewhat because of structural features such as faults and monoclines, the rocks are generally flat lying throughout the region. A **fault** is a fracture within rocks along which movement has occurred. In places, faults on the Colorado Plateau have created zones of weakness that have been eroded, forming steep, narrow canyons. The deep canyons, cut by running water, are indicative of the uplifting of the Colorado Plateau.

Downward erosion proceeds during uplift as running water attempts to reach lower elevations, and ultimately sea level. While the direction of flow of the Colorado River (which formed the Grand Canyon) has not been affected by faulting, the flow direction of some tributaries of the Colorado River are indeed fault controlled.

The extensive areas of exposed bedrock with limited vegetation and soil cover on the Colorado Plateau are a result of the region's arid to semi-arid climate. Colorful, brilliant rock layers are widely exposed. The rich colors of the Colorado Plateau result from oxidized iron (rust) present in the rocks, as well as surficial staining of exposed layers caused by **chemical weathering,** which is the process of the chemical alteration of minerals.

Much of the topography of the Colorado Plateau is steplike, reflecting differences in hardness of the rock layers. Hard, competent rock layers resistant to weathering and erosion, such as sandstone and limestone, form cliffs in the dry climate, while softer or fractured layers, such as shales, are more easily weathered and eroded, forming gentler slopes. **Sandstone** consists of cemented sand-sized fragments (mostly quartz), and **limestone** is composed mostly of crystalline fragments of the mineral calcite. **Shale** consists of silt and clay-sized fragments more susceptible to fracturing. For example, the Tonto Platform, a gently sloping shelf within the Grand Canyon, was formed through the weathering and erosion of the Bright Angel Shale. In places, resistant layers overlying more easily erodible layers have been undercut, creating shallow caves in which prehistoric cliff dwellings were constructed, such as at Navajo National Monument, east of the Grand Canyon.

The Grand Canyon, eroded by the Colorado River to one mile in depth and up to 18 miles wide, exposes horizontal Paleozoic-age (240 to 570 million years) sedimentary rocks down to just above the river, where older, Precambrian-aged (older than 570 million years) sedimentary, granitic, and metamorphic rocks are exposed. **Metamorphic rocks** are formed from pre-existing rocks that have been altered through an increase in temperature and/or pressure. The predominant metamorphic rock type in the Grand Canyon is **schist,** which is a coarse-grained foliated rock. **Foliation** occurs when platy or needle-shaped minerals in the rock align themselves in a parallel orientation in response to pressure exerted during metamorphism.

At the Grand Canyon, as at many areas on the earth's surface, breaks in the geologic sequence occur; that is, rocks of certain ages are missing. Such gaps are called **unconformities.** Unconformities are usually indicative of old erosional surfaces where missing rocks were eroded away, followed by a subsequent deposition of rocks younger than those that were eroded. Unlike the horizontal Paleozoic rocks at the Grand Canyon, the underlying Precambrian sedimentary rocks are tilted. Thus, an angular unconformity separates these two sequences of sedimentary rocks in the Grand Canyon. An **angular unconformity** is an unconformity in which the older strata dip more steeply than the overlying younger strata. This feature in the Grand Canyon is known as the "Great Unconformity," and it indicates that tilting and erosion took place between the deposition of the two sedimentary sequences. The tilted Precambrian sedimentary rocks rest upon older granite and intensely metamorphosed schist.

STRATIGRAPHY

Stratigraphy is the study of stratified, or layered rocks. As most of the Grand Canyon's walls are comprised of sedimentary rocks, this chapter describes the majority of rocks exposed there. In geologic publications we customarily describe the stratigraphy of a given area beginning with the oldest rocks. The reason for this procedure is the principle of superposition and the resulting interpretation of geologic history.

The principle of **superposition** states that, unless bedded rocks have been completely turned upside down (which is rare, but does occur), the rocks on the bottom of a sequence are the oldest and become progressively younger toward the top of the sequence. At the Grand Canyon, the oldest rocks occur at the bottom of the canyon. Thus, to interpret the geologic history of the Grand Canyon, we start with the oldest rocks, and work our way upward, or toward the youngest rocks.

In reality, the oldest rocks at the Grand Canyon are igneous and metamorphic, but since this chapter focuses on the stratified, or sedimentary rocks, those oldest rocks of the Grand Canyon are discussed in Chapter 3. In Chapter 7, the entire geologic history of the Grand Canyon is summarized.

In interpreting the geologic history of the Grand Canyon, it is important to understand the significance of the different depositional environments in which the sedimentary rocks were deposited. "The present is the key to the past" is a well-known phrase in geology. What this means is that geologic processes have remained constant over geologic time, so that processes observed today can be extrapolated back over geologic time. Thus, the depositional environments we observe today and the types of sediment deposited are similar to those in the geologic past. Therefore, through examination of sedimentary rocks, we can hypothesize about the depositional environments that existed at the time the rocks were formed. In each environment, a distinct sedimentary rock is formed with specific types of fossils and sedimentary structures. Table 2.1 shows different depositional environments and the types of sedimentary rocks formed within these environments, along with distinguishing characteristics.

The stratigraphic column shown in Figure 2.1 can help with gaining an understanding of the stratigraphy of the Grand Canyon. For those of you with a geologic map of the Grand Canyon, reviewing the map as you read this chapter can help you to understand the geology discussed herein.

TABLE 2.1
SEDIMENTARY ENVIRONMENTS OF DEPOSITION

ROCK TYPE	ENVIRONMENT	CHARACTERISTICS
• Conglomerate	River Channel	Rounded sand, gravel, cobbles, and boulders; no fossils or other structures
• Breccia	Alluvial Fan or Glacial Till	Angular gravel, cobbles, and boulders; no fossils or other structures
• Sandstone	Delta Beach Desert Dunes	Possible cross-bedding Possible shell fragments Cross-bedded, well sorted
• Shale	Floodplain Delta/Swamp Dry Lake Bed (Playa) Tidal Mud Flat	Possible mud cracks and fossils Possible fossils and interbedded coal Mud cracks, possible interbedded evaporites Ripple marks, marine fossils
• Limestone	Shallow to Deep Marine	Marine fossils; oolites
• Evaporites	Playa Shallow Marine	Interbedded shales Possible interbedded limestones
• Chert	Deep Marine	Possible marine fossils

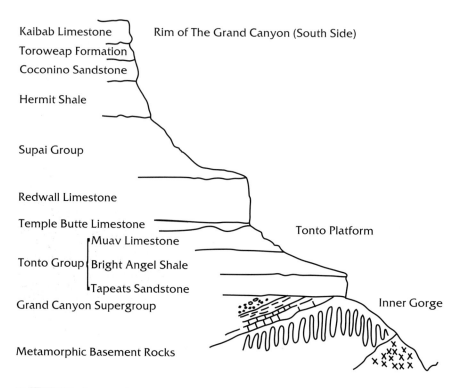

Figure 2.1 Stratigraphic column of the Grand Canyon. (From GEOLOGY OF NATIONAL PARKS by Harris et al. Copyright © 1997 by Kendall/Hunt Publishing Company. Reprinted by permission.)

PRECAMBRIAN ERA
Older Than 570 Million Years

The oldest sedimentary rocks in the Grand Canyon are Precambrian in age. Unlike the younger overlying Paleozoic rocks, the Precambrian bedded rocks are not horizontal, but are inclined as much as 15 degrees from horizontal. The Precambrian sedimentary rocks were deposited directly on even older igneous and metamorphic rocks. Based on the principle of **original horizontality,** we know that these sedimentary rocks originally

Figure 2.2 The Great Unconformity in the eastern Grand Canyon. Note beds dipping right (east) in the right–central portion of the photograph overlain by the horizontal, cliff-forming Tapeats Sandstone.

accumulated as flat layers. Some time after their formation, the beds were tilted and then eroded prior to deposition of the younger Paleozoic Rocks. This created the Great Unconformity discussed in Chapter 1 (Figure 2.2).

The Precambrian sedimentary rocks in the Grand Canyon are known collectively as the Grand Canyon Supergroup. Some basaltic lava flows and related bedded intrusive rocks called sills are also found in the Supergroup. **Basalt** is a type of volcanic rock that consists of iron and magnesium-rich minerals. A **sill** is a tabular igneous intrusion that has intruded parallel to the bedding of the host rock.

Everything about the Grand Canyon Supergroup is, well, super! It is estimated to range from 800 million to 1.2 billion years of age, and from 13,000 to 15,000 feet in thickness! Deposition accumulated in a large shallow depression known as the Apache Basin. Some of the better known formations comprising the Supergroup include, in descending order (from younger to older), the Chuar Group, Nankoweap Formation, Dox Sandstone, Shinumo Quartzite, Hakatai Shale, and Bass Limestone. A **group** is a continuous stratigraphic unit consisting of two or more formations. A **formation** is a distinguishable stratified rock unit of sufficient thickness and area to be mappable.

The Bass Limestone is the oldest and deepest sedimentary rock found at the Grand Canyon. Amazingly enough, fossils are present in the Bass Limestone, documenting the presence of life on earth one billion (yes, billion) years ago! The fossils present are those of primitive algae called **stromatolites** that represent the earliest colonial organisms that evolved in the ancient seas. The presence of quartzite, limestone, and shale in the Supergroup, common for rocks of this age, indicates that depositional environments similar to those of the present have existed since the Supergroup was formed.

PALEOZOIC ERA
240-570 Million Years

Sedimentary rocks at the Grand Canyon are mainly Paleozoic in age. A few minor remnants of younger, Mesozoic-age (63-240 million years) rocks are present near the Grand Canyon, but are generally insignificant and are not discussed here.

Tonto Group

530–550 Million Years Old
Part of the Cambrian Period

Three formations comprise the Tonto Group; from oldest to youngest they are the Tapeats Sandstone, Bright Angel Shale, and Muav Limestone. The significance of the Tonto Group is that this sequence of formations represents an excellent example of a **transgression,** or continual rise in sea level.

Formation Description

The Tapeats Sandstone, the oldest of the Paleozoic rocks present in the Grand Canyon, rests upon the Great Unconformity. This formation consists of brown, coarse-grained, cross-bedded sand, ranging in thickness from 100 to 300 feet. Due to the hardness of this rock, the Tapeats Sandstone forms steep cliffs.

16 Stratigraphy

The Bright Angel Shale is made up of 200 to 450 feet of green-colored **lithified,** or hardened, mud. When shale becomes lithified, it fractures. Hence, the Bright Angel Shale is easily eroded, which has resulted in the creation of the Tonto Platform described in Chapter 1 (Figure 2.3).

The mottled gray Muav Limestone ranges in thickness from 150 to 800 feet. The reason for the great variation in thickness is the uneven erosion that took place on the upper surface of the Muav Limestone. Where ancient rivers carved valleys in the Muav, prior to deposition of the overlying Temple Butte Limestone, the Muav is thin; the formation is thicker where former hills existed between the carved valleys.

Figure 2.3 View west of the Tonto Platform, which is incised by the Colorado River to create the inner gorge. Note the trail to Plateau Point in the lower central portion of the photograph.

Depositional Environments and Ancient Life

Together, the Tapeats Sandstone, Bright Angel Shale, and Muav Limestone display the record of a marine transgression. The Tapeats Sandstone represents the basal shoreline sand deposits that accumulated as marine waters transgressed from the west. As sea level continued to rise, and the shoreline migrated eastward, finer-grained mud was deposited over the former coastal sands, creating the Bright Angel Shale; gradational contacts with the Tapeats Sandstone and overlying Muav Limestone provide further evidence of transgression. Finally, the coastline had migrated so far eastward that fine-grained sediment deposition was replaced with carbonate deposition, forming the Muav Limestone.

Fossils are not found in the Tapeats Sandstone. However, trace fossils are present. **Trace fossils** are not actual remains of ancient life forms, but instead show evidence of their activities, such as tracks, trails, and burrows. The Tapeats Sandstone contains tubeworm burrows and trilobite trails. **Trilobites** were invertebrate animals, covered with a protective shell, which lived on the sea floor and moved about extracting nutrients from ingested sediment. The tubeworms also lived on the sea floor; Cambrian life existed only in marine environments.

Actual remains of worms, trilobites, and brachiopods are found in the Bright Angel Shale and Muav Limestone. **Brachiopods** were ancient shellfish related to present day mollusks. The reason fossil remains are present in the Bright Angel Shale and Muav Limestone, and not the

Tapeats Sandstone, is the environment of deposition; the Tapeats Sandstone was deposited in a coastal environment, where wave action destroyed any potential fossils. The Bright Angel Shale and Muav Limestone were deposited in an offshore, deeper water environment, below the reach of damaging wave action.

Temple Butte Limestone

360–370 Million Years Old
Part of the Devonian Period

Formation Description

The Temple Butte Limestone is composed mainly of dolomite, instead of calcite as most limestones are. **Dolomite** is a magnesium-rich calcite. The thickness of the Temple Butte Limestone is highly variable; it is absent where it was completely eroded, but is up to 1,000 feet thick in the western portion of the Grand Canyon.

A major unconformity exists between the Cambrian Muav Limestone and the overlying Devonian Temple Butte Limestone. Rocks of Ordovician and Silurian age are not present at the Grand Canyon. Thus, a gap of 160 million years (from 530 to 370 million years) exists, and we will never know geologically what occurred at the Grand Canyon during this span of geologic time. The highly eroded surface of the Muav Limestone suggests that any rocks of Ordovician and Silurian age were completely eroded from this portion of the Colorado Plateau.

The eroded surface of the Muav Limestone is also responsible for the variable thicknesses of both the Muav and Temple Butte Limestones. Because the top of the Temple Butte Limestone and the base of the Muav Limestone are generally horizontal, the Muav Limestone thickens as the Temple Butte Limestone thins, and vice versa. This relationship is demonstrated in Figure 2.4.

Depositional Environment and Ancient Life

The presence of conodonts reveals that the Temple Butte Limestone was deposited in a marine environment. **Conodonts** are microscopic marine-dwelling organisms. Also present are the remains of ancient fish. This is not surprising as the Devonian Period, when the Temple Butte Limestone formed, is known as "the age of fishes." This is significant, as fish were the first vertebrates to evolve on the earth.

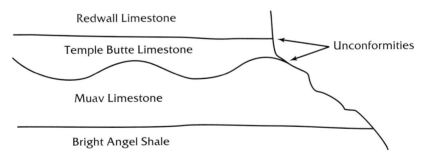

Figure 2.4   Relationship between Muav and Temple Butte Limestones.

Redwall Limestone

330–360 Million Years Old
Late Mississippian Period

Formation Description

The Redwall Limestone, ranging from 400 to 650 feet in thickness, is highly resistant to erosion and forms a distinct wall beneath the softer, overlying Supai Group (Figure 2.5). As the name implies, exposed cliffs of this formation are red; this red color is due to the staining of exposed surfaces from the chemical weathering of the overlying Supai Group, and is not inherent in the rock itself. On fresh, unexposed surfaces, the Redwall Limestone is gray in color.

Figure 2.5 Cliff-forming Redwall Limestone overlain by the less competent (more erodible) Supai Group.

Depositional Environment and Ancient Life

The Redwall Limestone contains fossil corals, which require warm, shallow seas for their growth. Other fossils present include brachiopods, crinoids, and bryozoans. **Crinoids** were similar to sea lilies, while **bryozoans** were similar to corals.

Supai Group

300–330 Million Years Old
Part of the Late Pennsylvanian and Early Permian Periods

Group Description

The Supai Group, which ranges in thickness from 600 to 950 feet, used to be considered a single formation. However, detailed studies of this unit revealed the existence of four distinct formations, and the name was revised as the Supai Group. Formations within the Supai Group consist of shales and cross-bedded sandstones rich red in color.

Depositional Environment and Ancient Life

Rocks of the Supai Group were deposited within a coastal environment and consist of overlapping flood-plain, swamp, and shallow-water deposits. The overlapping nature of these deposits indicates the occurrence of minor marine transgressions and regressions. The converse of transgressions, **regressions** represent a lowering of sea level. Plant fossils and reptile tracks identified within the Supai Group are significant because they represent the earliest signs of life on land that have been found at the Grand Canyon.

Hermit Shale

280–300 Million Years Old
Early Permian Period

Formation Description

The 300-foot thick Hermit Shale is composed of a lithified red silt. The rich red color of the Hermit Shale and underlying rocks of the Grand Canyon is the result of iron oxide present within the strata. The fractured nature of the Hermit Shale caused a relatively gentle slope to develop where the Hermit Shale is exposed, contrasting with the near vertical walls of the overlying, more resistant Coconino Sandstone (Figure 2.6).

Figure 2.6 An obvious geologic contact! Here, the white–colored Coconino Sandstone was deposited on the surface of the darker–colored Hermit Shale.

Depositional Environment and Ancient Life

The fine-grained nature of the Hermit Shale suggests that the environment of deposition was most likely shallow, quiet water in a lagoon, swamp, or flood-plain. Fossils in the Hermit Shale include those of plants and insects, and such trace fossils as worm trails and vertebrate tracks.

Coconino Sandstone

270–280 Million Years Old
Early Permian Period

Formation Description

Looking out over the Grand Canyon, the Coconino Sandstone forms the uppermost white band which can be seen along both the North and South Rims (Figure 2.7). The Coconino Sandstone is

Figure 2.7 Panorama of the North Rim of the Grand Canyon. Note how the light-colored Coconino Sandstone can be easily traced along the entire length of the photograph.

300 to 350 feet thick, and is composed of quartz sand that was deposited in massive, cross-bedded sand dunes.

Depositional Environment and Ancient Life

The preserved sand dunes indicate that the depositional environment was subaerial; that is, above water. Only trace fossils have been found in the form of reptile tracks; no actual fossils have been found in the Coconino Sandstone.

Toroweap Formation

260–270 Million Years Old
Middle Permian Period

Formation Description

The Toroweap Formation is made up of interbedded limestone and sandstone, and ranges in thickness from 200 to 250 feet. Its tan to reddish-tan color is slightly darker than that of the overlying Kaibab Limestone.

Depositional Environment and Ancient Life

The interbedded limestone and sandstone indicate a changing environment of deposition, in this case, an advance and retreat of a shallow sea. Fossils characteristic of a shallow sea are present, and include corals, brachiopods, bryozoans, and mollusks.

Kaibab Limestone

250–260 Million Years Old
Late Permian Period

Formation Description

The Kaibab Limestone, the youngest geologic formation present at the Grand Canyon, forms the surface of the Colorado Plateau in the immediate vicinity of the Grand Canyon. Thus, the Kaibab Limestone comprises the cap rock on both the North and South Rims of the Canyon. The Kaibab Limestone is light tan and ranges in thickness from about 300 to 350 feet.

Depositional Environment and Ancient Life

Fossils of mollusks, crinoids, brachiopods, sponges, bryozoans, and corals that are preserved in the Kaibab Limestone indicate this formation was deposited in a warm, shallow sea.

Younger Mesozoic-age rocks are absent in the immediate vicinity of the Grand Canyon. It's not that they weren't deposited; they were. However, erosional forces have removed all but a few remnants of these rocks. North and east of the Grand Canyon, in southern Utah and on the Navajo Indian Reservation, Mesozoic-age rocks are preserved, and it is within these rocks that dinosaur fossils occur. No dinosaur fossils are present in rocks at the Grand Canyon because these rocks predate the age of dinosaurs.

METAMORPHIC AND IGNEOUS ROCKS

Metamorphic and igneous rocks are exposed at the bottom of the Grand Canyon; Precambrian in age, they are the oldest rocks in the region. **Igneous rocks** are those which crystallize from magma, which is molten rock. Essentially all rocks in the earth's crust originated as igneous rocks; subsequent geologic processes transformed them into sedimentary or metamorphic rocks. Igneous rocks form either below the earth's surface as intrusive, or **plutonic** rocks, or on the earth's surface as extrusive, or **volcanic** rocks. Granite, a well-known and common type of plutonic rock, is rare at the Grand Canyon and is found only at certain locations within the deepest portion of the canyon. Volcanic rocks are also rare at the Grand Canyon, but an impressive **cinder cone,** a volcano comprised of erupted volcanic fragments, called Vulcan's Throne, erupted adjacent to the Colorado River in the western portion of the Grand Canyon just a little over one million years ago.

Metamorphic and Igneous Rocks

The metamorphic rocks make up the oldest rocks found at the Grand Canyon, and are estimated to have formed between 2 and 1.75 billion years ago. These ancient metamorphic rocks form the basement of the Grand Canyon. **Basement** rocks make up the oldest and deepest portion of the earth's crust, and are only exposed at the ground surface where younger rocks have been eroded away, such as in the Inner Gorge of the Grand Canyon. As the basement rocks of the Grand Canyon are extremely resistant, their erosion by the Colorado River has created the steep-walled, narrow canyon we call the Inner Gorge (Figure 3.1).

Based on the types of minerals comprising the metamorphic rocks at the Grand Canyon, and the corresponding temperatures and pressures necessary for their formation, it is estimated that these rocks formed at a depth as great as 13 miles below the earth's surface. The implication of this is that, between the time of their formation and their subsequent exposure prior to the beginning of the deposition of the Grand Canyon Supergroup about 1.2 billion years ago, about 13 miles of overlying rock were eroded. While 13 miles of erosion certainly sounds impressive, this erosion took place over a span of about 700 million years—quite a long time! In fact, this translates to an erosional rate of about one foot per 10,000 years. To put this rate of erosion into perspective, in Chapter 5 we compare this rate with the current rate of erosion of the Grand Canyon.

Metamorphic and Igneous Rocks 29

Figure 3.1 The Inner Gorge created by the Colorado River cutting through the very competent (hard) Precambrian metamorphic rocks found below the bedded sedimentary rocks at the Grand Canyon.

Three distinct metamorphic rock units have been identified at the Grand Canyon:

- Rama Schist and Gneiss
- Brahma Schist
- Vishnu Schist

To understand the differences between these rock units, some information about minerals would be helpful. **Minerals** are defined as homogeneous, naturally occurring, crystalline substances. What this means is that minerals have a distinct chemical composition (that is, a distinct combination of atoms of different elements) with atoms forming a regularly repeating internal structure (thus, crystalline). **Elements** are those substances found on the periodic table, such as hydrogen (H) and oxygen (O). As an example of how minerals form, hydrogen and oxygen combine at a ratio of 2:1, forming water (H_2O), a homogeneous substance. When water freezes in nature, the resulting ice is technically a mineral, as it is naturally occurring and crystalline! However, ice formed in your freezer is not a mineral, as it is not naturally occurring.

Rocks are naturally formed aggregates of one or more types of minerals bound together. Thus, it is important to make the distinction between rocks and minerals, as minerals comprise rocks. The significance of the differences in mineralogy among the metamorphic rock units of the Grand Canyon is that these differences are a result of different parent rocks; that is, the original rocks that existed prior to metamorphism.

The Rama Schist and Gneiss are composed primarily of the minerals quartz and feldspar, both light-colored minerals that lack iron and magnesium. A **gneiss** is a metamorphic rock exhibiting both foliated and massive textures, giving the rock a banded appearance. The chemistry and mineralogy of these rocks suggest that, prior to metamorphism, these rocks were volcanic in origin, with a mineralogy similar to that observed today.

The Brahma Schist is much darker than the Rama Schist and Gneiss due to a higher iron and magnesium content, resulting in the presence of dark-colored minerals such as amphibole and biotite (black mica). The chemistry and mineralogy of the Brahma Schist make it likely that the Brahma Schist formed through the metamorphism of basaltic-type igneous rocks.

The Vishnu Schist, the best known of the metamorphic rocks at the Grand Canyon, is composed of both dark and light-colored minerals, including quartz, muscovite (white mica), garnet, and amphibole. Because of this mixture of dark and light-colored minerals, it is probable that the original rocks were fine-grained sedimentary rocks with interbedded lava flows.

The intense temperatures and pressures necessary for the formation of these metamorphic rocks must have caused at least some melting to occur during metamorphism. The results of this partial melting can be observed near the Colorado River in the form of migmatites (Figure 3.2). **Migmatites,** which are thin layers, veins, or lenses of coarse-grained granitic-type rocks found within intensely metamorphosed rocks, form through the partial melting of rocks undergoing metamorphism.

Shortly (geologically speaking!) after metamorphism took place, the rocks were intruded by granitic-type rocks. As the granitic rocks are not metamorphosed, we know that they must be younger than the

Figure 3.2 Migmatite characteristic of the Vishnu Schist.

metamorphic rocks. The granitic rocks are sometimes known as the Zoroaster Granite, but recent studies have identified up to three distinct igneous intrusive rock types.

The deep erosion of the Grand Canyon by the Colorado River provides a window into the underlying basement rocks. The exposure of these rocks gives us a glimpse of continental crust that is covered by younger sedimentary rocks throughout most of the Colorado Plateau, as well as the rest of North America and the world. Continental crust ranges from about 20 to 25 miles in thickness. Most of this thickness is comprised of metamorphic and plutonic rocks; sedimentary rocks occur only as a relatively thin (geologically speaking) skin on the outermost portion of the crust. Sedimentary rocks are not found deeper in the earth's crust, because the increased temperatures and pressures either metamorphose or melt the sedimentary rocks, changing their original sedimentary nature forever.

STRUCTURAL GEOLOGY

At the Grand Canyon, the Paleozoic sedimentary rocks are nearly horizontal, or flat-lying, like a layer cake or stack of pancakes. Here we might not expect to see structural features, such as folds and faults (a **fold** is the bending of layered rocks). However, both exist at the Grand Canyon and while they have not significantly deformed or warped the rocks, they have created features and landscapes of enough interest that it would be remiss not to describe them.

As discussed in Chapter 1, the Colorado Plateau (all 150,000 square miles) has been uplifted; the plateau surface ranges from 5,000 to 7,000 feet above sea level. This uplift, as well as creation of all geologic provinces, is a result of plate tectonic activity. The uplifting of such a large area could not take place without the creation of faults because uplift pressures could not be constant throughout the entire plateau region. Going back to our layer cake analogy, more often than not, just transferring a cake from one plate to another may result in the

36   Structural Geology

breaking, or "faulting" of the layers. This happens because the cake is not strong enough to withstand stresses placed on the layers being moved. The same thing has happened at the Colorado Plateau. As strong as the rocks are, they are not strong enough to withstand the stress of uplift, causing faulting to occur.

A map of some of the better known structural features at the Grand Canyon is presented in Figure 4.1. This map does not show all of the faults

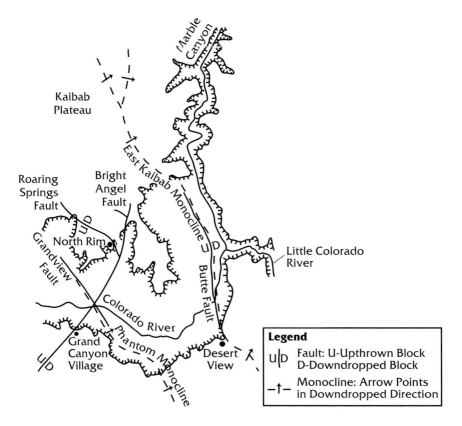

Figure 4.1   Major structural features of the Grand Canyon.

and folds present at the Grand Canyon, but does identify the structural features described in this chapter.

The Bright Angel Fault is the best known fault at the Grand Canyon. This northeast-trending fault cuts through the west end of Grand Canyon Village on the South Rim, and extends northeast beyond the Colorado River into Bright Angel Canyon, the straight, steep canyon from which Bright Angel Creek empties into the Colorado River at Phantom Ranch. The Bright Angel Trail, which connects the South Rim with Indian Gardens, approximately parallels this fault. In fact, the Bright Angel Fault created a linear zone of weakness susceptible to erosion by Bright Angel Creek. This is true for many of the faults in the canyon, but not all of them. Some canyons were eroded along fault lines, while others were not. The trace of the Colorado River is not affected by faulting.

The Roaring Springs Fault trends northwest from the Bright Angel Fault north of the Colorado River. This fault has created a side canyon to Bright Angel Canyon. Of significance is the termination of this fault at the Bright Angel Fault. What has happened to this fault southeast of the Bright Angel Fault? We don't know for certain, but perhaps some other faults southeast of the Bright Angel Fault were once a continuation of this fault.

The principle of crosscutting relationships can be used to evaluate the relative ages of these faults to one another. This principle states that geologic features, in this case faults, are *older* than the features cutting

38   Structural Geology

across or offsetting them, such as other faults (Figure 4.2). Thus, we can surmise that the Bright Angel Fault is younger than the Roaring Springs Fault, since it cuts across and offsets this fault, causing it to terminate at the Bright Angel Fault.

The most prevalent type of fold present at the Grand Canyon is the monocline. **Monoclines** are folds occurring in horizontally bedded rocks that bend in one direction only, resulting in identical beds being found at different elevations on either side of the monocline. Two major monoclines

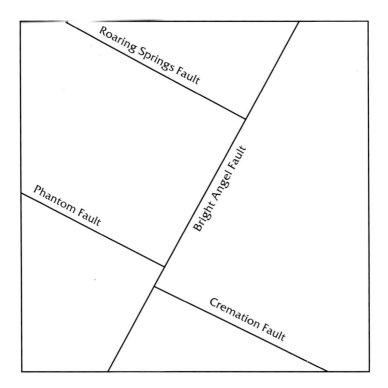

Figure 4.2   Schematic representation of cross–cutting relationships in the Grand Canyon. Bright Angel Fault is younger than Cremation, Phantom, and Roaring Springs Faults because the latter two are truncated by the Bright Angel Fault.

present at the Grand Canyon are the Grandview-Phantom and East Kaibab Monoclines (Figure 4.1).

How have monoclines formed at the Grand Canyon? Prevailing thought suggests that monoclines at the Grand Canyon have formed through the renewed vertical movement of older faults in the underlying Precambrian rocks that are now covered over by the Paleozoic sedimentary rocks. This is demonstrated in Figure 4.3. In fact, it now appears that after creation of the monoclines, continued underlying fault movement resulted in the faulting of the overlying Paleozoic rocks. That is, the Paleozoic rocks were folded into the monoclines we see today, but subsequent fault movement proved too great for the Paleozoic rocks, resulting

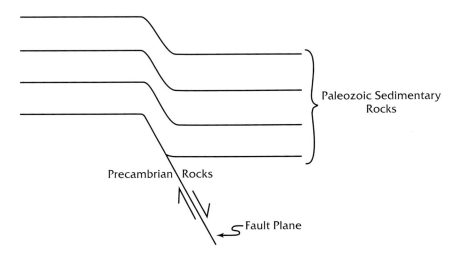

Figure 4.3 Origin of monoclines at the Grand Canyon.

in their being faulted as well. Evidence for this is the presence of significant faults parallel and adjacent to the two major monoclines (Figure 4.1):

- Grandview-Phantom Monocline and Fault
- East Kaibab Monocline and Butte Fault

One feature the East Kaibab and Grandview-Phantom Monoclines have in common is that they are east-facing; that is, the east side has been down dropped relative to the west side. This suggests a regional structural pattern, and that formation of these monoclines and corresponding faults is related.

The East Kaibab Monocline, the largest of the monoclines, is responsible for creation of the Kaibab Plateau on the North Rim, forming the eastern boundary of the Kaibab Plateau. As noted earlier, the top of the Kaibab Limestone is the uppermost surface of most of the Colorado Plateau in the vicinity of the Grand Canyon. However, on the Kaibab Plateau, a six-mile long ridge of Redwall Limestone is exposed at the ground surface along the axis of the north-south trending East Kaibab Monocline. The East Kaibab Monocline is responsible for the uplifting of the Redwall Limestone at this location. As the Kaibab and Redwall Limestones are separated by 1,500 to 2,000 feet of sandwiched formations (Toroweap Formation, Coconino Sandstone, Hermit Shale, and Supai Group), we know that at least this amount of vertical movement has occurred along the East Kaibab Monocline (Figure 4.4). As a result of this uplift, the elevation of the Kaibab Plateau on the North Rim is about 8,000 feet above sea level, about 1,000 feet higher than the South Rim, which has an elevation of about 7,000 feet.

Structural Geology 41

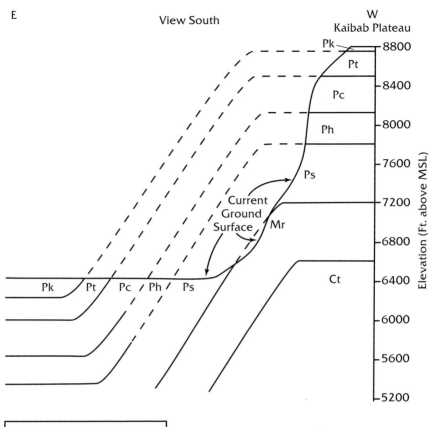

Legend
Pk - Kaibab Limestone
Pt - Toroweap Formation
Pc - Coconino Sandstone
Ph - Hermit Shale
Ps - Supai Group
Mr - Redwall Limestone
Ct - Tonto Group

Figure 4.4 Schematic cross-section across the East Kaibab Monocline.

GEOMORPHOLOGY

Geomorphology addresses the configuration of the earth's surface and the changes that take place in the evolution of landforms. Landforms on the Colorado Plateau and at the Grand Canyon such as buttes, pinnacles, and spires are unique, and occur as a result of the combination of arid climate, horizontal layering, and differing formation hardness.

The arid climate results in the absence of both soil and vegetation. Unlike in moist climates where bedrock is covered by soil and vegetation, rocks are exposed. As the exposed formations on the Colorado Plateau are flat lying, outcrop patterns follow surface contours. And as the formations exhibit varying degrees of hardness, they are subject to differential weathering. **Differential weathering** is the irregular rate of weathering caused by differences in rock hardness. The differences in hardness of vertically adjoining formations results in the creation of the cliff and slope-forming topography discussed in Chapter 1. Cliffs slowly retreat as the weathering and erosion of underlying softer formations undermines them. Large blocks of rock break off as rock falls, and accumulate at the base of cliffs as talus slopes. This retreat occurs at differing rates along the cliff face, primarily as a result of drainage patterns; retreat is quicker in drainages where more water is channeled. It is this irregular retreat over time that ultimately results in the creation of resistant buttes, pinnacles, and spires (Figure 5.1). These features are the last erosional remnants of continuously-bedded formations. And ultimately, these stately features will themselves become weathered and eroded.

Anyone who has been fortunate enough to raft down the Colorado River in the Grand Canyon has experienced the many rapids present

Figure 5.1 "Temples" in the Grand Canyon. These white-capped erosional remnants exist due to the hardness of the Coconino Sandstone (the white formation). Once the Coconino Sandstone erodes completely, the temples will erode at a more accelerated rate.

along this stretch of the river. In between the rapids are stretches of quiet water, where one can catch one's breath between rapids. While rafting through a rapid, most people don't stop to consider why the rapid is located where it is; they're either terrified or having too much fun. But for us geologists, we never stop thinking about geology!

Rapids are located where tributaries empty into the Colorado River. Debris flows, an episodic type of flash flood, are initiated by slope failures that occur during intense rainfall, and transport boulders and sediment down tributaries, depositing material in the Colorado River channel. Certainly, the Colorado River flushes the finer-grained debris (sand, silt) downstream. The coarse boulders remain and accumulate, requiring the water of the Colorado River to flow over and around these boulders, creating the rapids (Figure 5.2). And the rapids themselves are currently growing as a result of man's activities on the Colorado River. Upstream construction of Glen Canyon Dam completed in 1963 lowered the magnitude of periodic Colorado River flooding, reducing the ability of the river to transport larger boulders. As a result, since 1984, debris flows have created four new rapids and enlarged 17 others!

How exactly has the Colorado River created the Grand Canyon? And how long has the river taken to create the Canyon? In as much as the Colorado River is responsible for the creation of the Grand Canyon, let us first look at the evolution of rivers in general.

Figure 5.2 Colorado River rapids created by the deposition of large debris (boulders) from a tributary. Look closely, and you can see the author getting soaked!

As water runs downhill to the topographically lowest point or points, it makes sense that in areas of relief, rivers are found in the topographically lowest locations. But which came first, rivers or the valleys in which they are found? At first glance, this appears to be a chicken and egg discussion. But upon further reflection, what makes sense is that they're inextricably linked.

Tectonic uplift and rainfall have occurred simultaneously throughout geologic time. Thus, as an area becomes uplifted, rain falling on the uplifted areas becomes surface runoff and flows downhill, beginning the erosional process. Remember that if the uplifted area was totally flat with no relief, and the uplift occurred evenly over a large area, the runoff would have nowhere to flow, and would collect locally. In nature, few land surfaces are flat enough to collect runoff. These surfaces are called **playas,** which are lakebeds found in arid areas that are dry throughout most of the year. However, as we learned in the previous chapter, uplift does not occur evenly; faults are created, and the rate of uplift varies on either side of the fault. Thus, geologic structures are responsible, in part, for creating topographic variations on the earth's surface. These variations, combined with the effects of rainfall and runoff, bring about the development of rivers, drainages, and other topographic features.

Various theories have been proposed over the years for the origin of the Colorado River and the Grand Canyon. The generally accepted theory starts with an understanding of what has happened to the area now known as the Colorado Plateau beginning about 20 million years ago. The structural history of the area has had a major effect on the development of drainage.

Deposits of ancient gravels, known as "Rim gravels," are found on the surface of the southern and southwestern portions of the Colorado Plateau. The Rim gravels contain rock types such as granite and gneiss that do not occur on the surface of the Colorado Plateau, but are found to the south and southwest. Thus, these gravels had to have originated from what were then topographically higher locations. Similar gravels have also been identified north of the western portion of the Grand

Canyon. Thus, we can deduce that the Grand Canyon could not have existed when the gravels were deposited, as the Grand Canyon would have acted as a barrier, preventing their transport and subsequent deposition to the north. The Rim gravels are overlain by a volcanic formation (Peach Springs Tuff) that has been age-dated at 17 to 20 million years. The Peach Springs Tuff was volcanic ash that originated from an area southwest of the Grand Canyon, and flowed northeastward. Thus, we know that the Grand Canyon is younger than these dated volcanic rocks.

Starting about 17 million years ago, areas surrounding the Colorado Plateau to the northwest, west, and south were affected by faulting that eventually formed the Basin and Range Province. Pre-existing drainage onto the Plateau was destroyed, and the geologic provinces began to develop as we see them today. However, even as recently as 5 to 6 million years ago, it does not appear that erosion of the Grand Canyon had begun. This conclusion is based on the presence of interbedded limestones and volcanic rocks abutting the western boundary of the Colorado Plateau, where the present day Colorado River flows into Lake Mead, which was man-made by the construction of Hoover Dam. These rocks were deposited in a basin bounded on the east by the Colorado Plateau. The occurrence of limestone indicates the presence of a lake as recently as 5 to 6 million years ago based on the age-dating of interbedded volcanic rocks. Had the Colorado River existed at that time and emptied into the lake, sediments carried by the river would have been deposited. However, no such sediments have been found, indicating that the Colorado River could not have existed at that time.

About 5.5 million years ago, tectonic forces along the Arizona-California border (the same forces responsible for the San Andreas Fault in California) created a through-flowing drainage along the area now known as the lower Colorado River, which forms the border between Arizona and California. River-deposited sediments overlie volcanic rocks that have been dated at 5.3 million years; thus, this is the oldest age that these sediments can be. Within these same sediments in extreme southeastern California, small fossils found only in the Cretaceous Mancos Shale (which overlies the Kaibab Limestone on the Colorado Plateau but is not found in the immediate vicinity of the Grand Canyon due to its removal by erosive forces) were discovered. Thus, it can be deduced that the Colorado River began flowing through the Colorado Plateau and creating the Grand Canyon about 5.3 million years ago.

About 3.8 million years ago (again, based on age-dating), a basaltic eruption flowed down a tributary of the Colorado River in the northwestern portion of the Grand Canyon. This basalt is now located about 300 feet above the present day Colorado River. This indicates that the Colorado River had already eroded down to near its present depth by 3.8 million years ago. Another basaltic eruption, Vulcan's Throne, occurred just over one million years ago, and these lavas cascaded to the present elevation of the Colorado River from the rim, indicating that minimal downcutting of the Grand Canyon has taken place in the last million years.

The conclusion to be drawn from this information is that the Grand Canyon was eroded in, geologically, a very short time; four million years or less! Given the resistance to erosion of most of the formations present at the Grand Canyon, the rate of widening of the Grand Canyon is small in comparison to the rate of downcutting. The steep canyon walls are not typical of how most rivers erode and evolve.

The Grand Canyon averages about one mile (5,280 feet) in depth. To calculate the rate of downcutting using the dates described above, it appears that the Grand Canyon was eroded to within 300 feet of its current depth between 5.3 and 3.8 million years ago, a span of 1.5 million years. Thus, it appears that about 5,000 feet of downcutting took place over this period of time, an average rate of downcutting of 0.003 feet per year, or 30 feet per 10,000 years. Remember, in Chapter 3, we calculated an erosional rate of one foot per 10,000 years for the exposure of the metamorphic rocks of the Grand Canyon back in the Precambrian. Thus, the recent rate of erosion at the Grand Canyon has been about 30 times faster!

Even an erosional rate of 30 feet per 10,000 years may not sound like much, but considering the average human life span is about 70 years, during our grandparent's lifetime, the Colorado River deepened the Grand Canyon by almost one-fourth of a foot. Puny, perhaps, by human standards, but geologically, this rate of erosion is phenomenal! It remains to be seen whether this rate of erosion will continue. With the construction of Glen Canyon Dam upstream from the Grand Canyon, man has now altered the flow of the Colorado River through the Grand Canyon.

GROUNDWATER AND SPRINGS

In the previous chapter, we learned about the effect of tributaries on the creation of rapids in the Colorado River. But what about the tributaries themselves? What is their origin? In this chapter, we will learn about groundwater at the Grand Canyon and the origin of springs, and thus of tributaries of the Colorado River.

Groundwater is water that occurs in the subsurface under saturated conditions; that is, will flow into a well or other opening. The top of the zone of saturation is the **groundwater table.** Many people think of groundwater as occurring as underground rivers. This is mostly erroneous. While a very small percentage of groundwater does occur as underground rivers, most groundwater occurs in pore spaces of granular rocks (primarily sedimentary) or in rock fractures. Underground rivers form when slightly acidic groundwater flows through limestone. Calcite comprising the limestone begins to dissolve when in contact with slightly acidic groundwater. This dissolution, over geologic time, creates cavities where water flows and thus underground rivers are formed. Again, contrary to popular belief, underground rivers are rare.

What causes groundwater to flow? We know surface water flows downhill as a result of gravity. Groundwater flows under the same concept. Because we cannot see groundwater (as it is underground!), the fact that groundwater flows may be a difficult concept to understand. But groundwater, like surface water, flows from higher to lower elevations; albeit much slower. It is important to understand that, in many instances, groundwater flow mimics surface topography. But groundwater flow doesn't always follow surface topography, and in some instances may flow in a different direction.

When the groundwater table intersects surface topography (usually through a rapid drop in surface topography, such as a cliff), springs are formed. **Springs** are groundwater flowing onto the ground surface. Think of springs as transition zones where groundwater flows to the ground surface, creating a creek; it is important to understand that ground and surface water are related, and part of a single hydrologic system. When a creek flows into another creek or river, it becomes a tributary. This is how tributaries to the Colorado River in the Grand Canyon have formed. At the head of essentially every tributary of the Colorado River in the Grand Canyon are one or more springs of varying size. Some of these springs are large, creating perennial (year round) surface water flow, while others are miniscule and little more than a trickle. Surface water flow in the tributaries is the accumulated flow of multiple springs that issue from the sides or bottoms of the tributaries.

Flowing surface water originating at springs causes **headward erosion;** that is, the back cutting of a drainage, increasing its length over geologic time. Thus, over time, springs migrate in an upstream direction as a result of the erosive forces they create.

Groundwater originates as precipitation in the form of rain or snow. As rainfall (or melted snow) flows on the ground, a portion percolates into the soil. Due to the force of gravity, the water is drawn vertically downwards (in an unsaturated condition) until it either reaches the groundwater table or an impervious layer. When the water reaches the groundwater table, it is said to have recharged the water table. If an impervious layer lies above the groundwater table, the percolating water becomes "perched" on this layer, and if enough water is present to create saturated conditions, a zone of perched groundwater develops. This is exactly what has happened at the Grand Canyon.

Precipitation at the North and South Rims infiltrates the permeable Kaibab Limestone, which caps both rims. **Permeability** is the ability of a formation to transmit groundwater. The percolating water continues to move downwards until perching on impermeable formations occurs. Once perched, groundwater begins moving in a horizontal manner until reaching the walls of the Grand Canyon. The groundwater then flows onto the ground surface as a spring. At the Grand Canyon, groundwater above the Redwall and Muav Limestones has, for the most part, drained downward to these formations, where it moves horizontally along joints and solution channels towards springs. These permeable formations are responsible for the major springs at the Grand Canyon.

Four springs of significance at the Grand Canyon are discussed herein, in order from higher to lower elevations. Veseys Paradise, a major spring, was discovered in 1869 by John Wesley Powell's expedition down the Colorado River, the first exploration of the Colorado River. Powell's group turned a sharp corner to see a beautiful canyon wall with fountains of water bursting from the Redwall Limestone. Beneath the spring, moist walls are covered with lush green vegetation, in stark contrast to adjacent exposed rock (Figure 6.1).

The following three springs originate in the Muav Limestone, which is underlain by the Bright Angel Shale, a relatively impermeable formation causing the creation of springs at the base of the Muav. Roaring Springs is located near the upstream end of Bright Angel Creek. These springs are easily visible hiking into the Grand Canyon from the North Rim (Figure 6.2). Powell did not discover Roaring Springs, but did discover Bright Angel Creek where it flows into the Colorado River at Phantom Ranch. Interestingly, he originally named it Silver Creek, but later changed the name to Bright Angel to compensate for having named another tributary Dirty Devil!

Groundwater and Springs 57

Figure 6.1 Veseys Paradise. This beautiful spring can only be observed (and enjoyed) by rafting down the Colorado River from Lees Ferry, near Utah.

58 Groundwater and Springs

Figure 6.2 Roaring Springs. This prolific spring supplies water to both rims, and can be viewed by descending the Bright Angel trail from the North Rim.

Roaring Springs contribute most of the base flow to Bright Angel Creek. Roaring Springs issues from solution channels high on a steep slope near the base of the Muav Limestone. The springs cascade several hundred feet before reaching the creek bed. Roaring Springs discharges so much water that they are now the primary source of water for both the North and South Rims (with the aid of pumps and piping). Thus, if you use a water fountain at either rim, you are drinking water from Roaring Springs.

The well-known Indian Garden along the Bright Angel trail from the South Rim is also a spring. Prior to the development of Roaring Springs, Indian Garden Springs provided water for the South Rim. However, the yield of this spring was insufficient to supply the demand of the South Rim with the increasing popularity of Grand Canyon National Park.

Havasu Spring is located west of the National Park boundary, and is the major source of water for Havasu Creek. The chemistry of the water from Havasu Spring differs from the chemistry of the springs previously discussed. Water from Havasu Spring is higher in calcium and carbonate. And, as you'll recall from previous chapters, these are the components of the mineral calcite.

As a result of increased calcium and carbonate, once discharged, water from Havasu Spring precipitates calcite along Havasu Creek. This calcite is precipitated in the form of travertine. **Travertine** is a type of limestone deposited in fresh (as opposed to salt) water. Travertine is also found at several other locations at the Grand Canyon. The origin of the calcium and carbonate in the spring water at these locations is the limestone present at the Grand Canyon. Groundwater feeding the springs where travertine is deposited was originally slightly acidic. As the groundwater flowed through the various limestone formations, it slowly dissolved some of the limestone, bringing calcium and carbonate into solution and increasing their concentrations in the groundwater. Once the groundwater emanated from springs, differing environmental conditions above ground caused the calcium and carbonate to precipitate, creating the travertine deposits.

GEOLOGIC HISTORY

Rocks at the Grand Canyon, spanning the last two billion years of geologic time, provide a history, albeit incomplete, of northern Arizona.

PRECAMBRIAN ERA
Older Than 570 Million Years

- At some time prior to two billion years ago, volcanic rocks rich in quartz and feldspar were erupted. These were followed by the eruption of darker volcanic rocks, basaltic in composition. These basaltic eruptions probably occurred under water. Subsequently, thick sequences of fine-grained sediment were deposited with intermittent lava flows.
- About two billion years ago, the entire sequence of rocks was buried to a depth of up to 13 miles below the earth's surface. This deep burial resulted in intense regional metamorphism of these rocks, creating the rocks we see today at the bottom of the Grand Canyon. How did these rocks get buried so deeply? Perhaps they

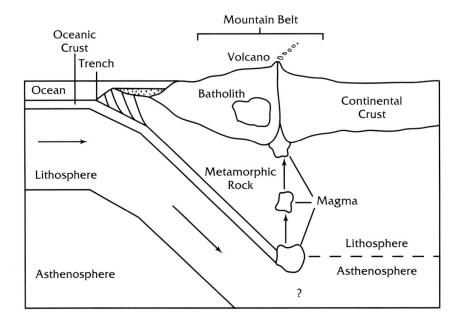

Figure 7.1 Schematic cross-section of a subduction zone. Dense oceanic crust composed of basalt and overlying sediments is thrust beneath less dense continental crust when they collide as a result of plate tectonics. Increased temperature and pressure at depth metamorphose subducted oceanic crust into metamorphic rocks.

were subducted as a result of plate tectonics (Figure 7.1). The suite of basalts overlain by fine-grained sediments would be consistent with our knowledge of oceanic crust. At about this point in geologic time, North America collided with oceanic crust to the south, creating an east-west trending mountain belt known as the Yavapai-Mazatzal orogen in what is now Arizona (Figure 7.2). This **orogen,** or zone of deformed, metamorphic rock, increased the size of the North American continent.

Geologic History 63

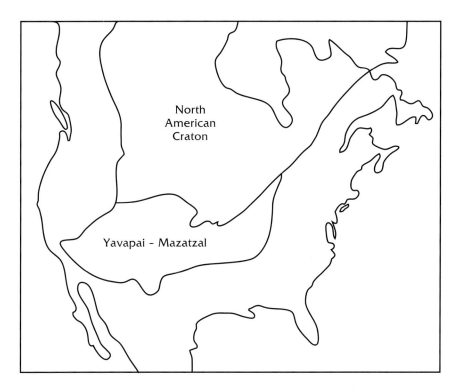

Figure 7.2 Precambrian evolution of North America (taken from Wicander and Monroe, HISTORICAL GEOLOGY). The collision of the early North American continent with oceanic crust to the south created the Yavapai–Mazatzal orogen in what is now Arizona, increasing the size of the continent. The schematic cross-section shown in Figure 7.1 is representative of such a collision.

- About 1.75 billion years ago, the now metamorphosed rocks were intruded by quartz-rich magma that crystallized into granitic rock. Prior to about 1.2 billion years ago, the entire sequence of granitic and metamorphic rocks was uplifted above sea level and exposed by the erosion of overlying rocks.

- By about 1.2 billion years ago, the granitic and metamorphic rocks were covered by water, and for the next 400 million years, sedimentary rocks now known as the Grand Canyon Supergroup were deposited in a depression known as the Apache Basin (Figure 7.3). The depositional environments that existed at the time were similar to those of today. At least 13,000 feet of sedimentary rocks, and probably much more than that were deposited. Some volcanism also took place, resulting in interbedded volcanic rocks and related igneous intrusives.

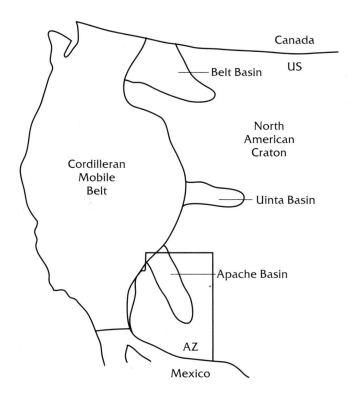

Figure 7.3 Late Precambrian sedimentary basins in the western United States. The Grand Canyon Supergroup was deposited within the Apache Basin in what is now northern Arizona (taken from Wicander and Monroe, HISTORICAL GEOLOGY).

Geologic History

- Between 800 and 550 million years ago, the Grand Canyon Supergroup and underlying granitic and metamorphic rocks were faulted and tilted. Deformation was not severe, but this tectonic movement was sufficient to lift these rocks above sea level, creating an erosional surface now known as the "Great Unconformity" (Figure 7.4).

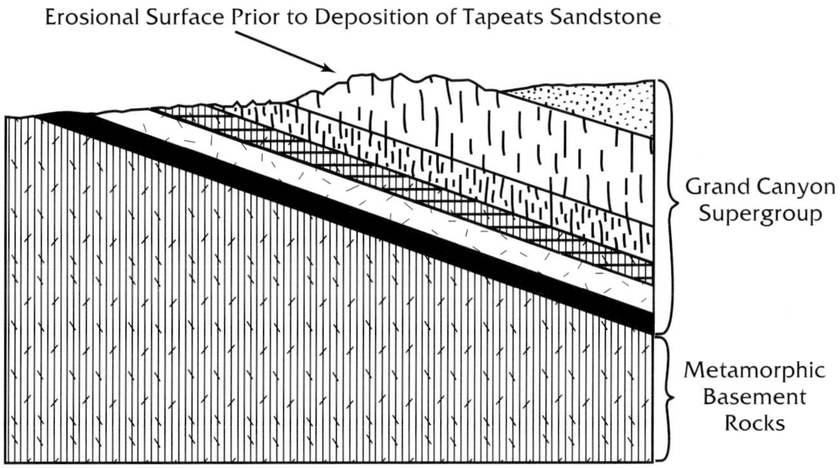

Figure 7.4 Cross-section of vicinity of the Grand Canyon prior to deposition of the Tapeats Sandstone. The erosional surface shown here is called the "Great Unconformity."

PALEOZOIC ERA
240-570 Million Years

- During the Cambrian Period, 550 million years ago, northern Arizona was in a coastal environment in which the Tapeats Sandstone accumulated. As transgressing waters deepened, the Bright Angel Shale was deposited, followed by the Muav Limestone 530 million years ago.

- A gap in the geologic record brought about another unconformity, between 530 and 370 million years ago. Rocks of Ordovician and Silurian age are not present at the Grand Canyon. Whether they were never deposited, or were eroded shortly after their deposition, we may never know. However, we do know that at some point during this span of geologic time, the upper portion of the Muav Limestone was lifted above sea level and eroded; this is evident from the highly irregular upper surface of the Muav Limestone.
- During the Devonian Period 360-370 million years ago, the upper surface of the Muav Limestone was covered by water, and the Temple Butte Limestone was deposited.
- A minor unconformity is present at the top of the Temple Butte Limestone. However, northern Arizona again became submerged during the Mississippian Period 330-360 million years ago, and the Redwall Limestone was deposited.
- Another minor unconformity is present at the top of the Redwall Limestone. As the Redwall Limestone began to be submerged during the Pennsylvanian/Permian Period 300-330 million years ago, deposition of the first of the four formations comprising the Supai Group began. Rocks of the Supai Group indicate that deposition occurred within a coastal environment consisting of flood-plains, swamps, and shallow coastal waters that alternately transgressed and regressed over the period of deposition.

- Sea level fell and rose yet again, creating another minor unconformity at the top of the Supai Group. During the Permian Period, 280-300 million years ago, the Hermit Shale accumulated in a depositional environment that must have been somewhat similar to that existing earlier in the Permian when upper formations of the Supai Group were deposited.
- Still during the Permian Period, 270-280 million years ago, sea level fell slightly, exposing coastal sands. Prevailing winds created great sand dunes that lithified into the Coconino Sandstone.
- As sea level rose in late Permian time, 260-270 million years ago, interbedded limestones and sandstones were deposited, creating the Toroweap Formation.
- A drop in sea level created a minor unconformity at the top of the Toroweap Formation. Sea level then rose in the late Permian 250-260 million years ago, and the Kaibab Limestone was deposited, which marks the end of the Paleozoic Era at the Grand Canyon.

MESOZOIC ERA
65-240 Million Years

- Additional deposition took place during the Mesozoic Era, which ranged from 65 to 245 million years ago. However, these rocks were almost completely eroded in the immediate vicinity of the Grand Canyon between about 20 and 65 million years ago.

CENOZOIC ERA
Younger Than 65 Million Years

- About 20 million years ago, the Rim gravels were deposited on the surface of the western portion of what is now the Colorado Plateau. The Peach Springs Tuff was erupted in this same area between 17 and 20 million years ago, covering the Rim gravels.
- About 17 million years ago, faulting along the northwest, west, and southern margins of the Grand Canyon began the creation of the Basin and Range province. Concurrent regional uplift of what is now the Colorado Plateau reactivated old, Precambrian-age faults, creating the monoclines and younger faults present at the Grand Canyon.
- The ancestral Colorado River began to erode the Grand Canyon about 5.3 million years ago. Most of the canyon had been carved by about 3.8 million years ago. Figure 7.5 shows a geologic cross-section of the Grand Canyon as it appears today.
- Concurrent with erosion by the Colorado River, tributary erosion took place, widening the Grand Canyon. Tributary erosion proceeded as a result of headward erosion by springs.
- Vulcan's Throne and associated basaltic lava flows erupted 1.1 million years ago in the western portion of the Grand Canyon.
- Flow in the Colorado River and adjoining tributaries continues to this day, providing water and recreation to visitors of the Grand Canyon.

Geologic History 69

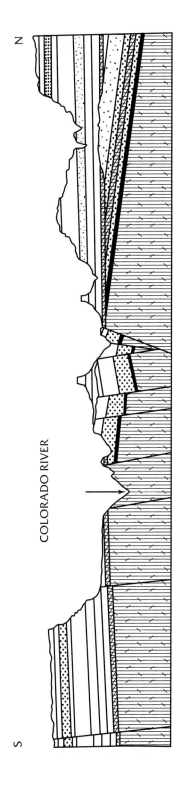

Figure 7.5 Schematic geologic cross-section of the Grand Canyon (adapted from Grand Canyon Natural History Association).

GLOSSARY

Angular Unconformity — An unconformity in which the older strata dip steeper than the overlying younger strata.

Basalt — A dark colored volcanic rock that consists of iron and magnesium-rich minerals.

Basement Rock — Rock comprising the oldest and deepest portion of the earth's crust, and is only exposed at the ground surface where younger rocks have been eroded away.

Brachiopods — Marine shellfish found in Paleozoic rocks similar to mollusks.

Bryozoans — Ancient sea-floor life similar to corals.

Calcite — A mineral comprised of calcium carbonate; a major component of limestone.

Chemical Weathering — The process of the chemical alteration of minerals.

Cinder Cone — A volcano comprised of erupted volcanic fragments; usually associated with basalt.

Conodonts — Microscopic marine-dwelling organisms.

Crinoids — Ancient sea-floor life similar to sea lilies.

Differential Weathering — Irregular rate of weathering caused by differences in rock hardness.

Dolomite — A magnesium-rich calcite.

Element — Substance found on the periodic table.

Fault — A fracture within rocks along which movement has occurred.

Fold — The bending of layered rock.

Foliation — Metamorphic texture formed when platy or needle-shaped minerals align themselves in a parallel orientation in response to pressure.

Formation — A distinguishable stratified rock unit of sufficient thickness and area to be mappable.

Fossil — Remains of ancient life preserved in sedimentary rock.

Geologic Province — A large region characterized by similar geologic history and development.

Geomorphology — The configuration of the earth's surface and the changes that take place in the evolution of landforms.

Gneiss — Metamorphic rock containing both foliated and massive textures, giving the rock a banded appearance.

Groundwater — Water that occurs in the subsurface under saturated conditions; that is, will flow into a well or other opening.

Groundwater Table — The top of the zone of saturation.

Group — A major stratigraphic unit consisting of two or more formations.

Headward Erosion — The back cutting of a drainage, increasing its length over geologic time.

Igneous Rock — Rock which forms from the cooling of molten rock.

Limestone — Sedimentary rock composed primarily of the mineral calcite.

Lithified — The process of the hardening of sediments into sedimentary rock.

Metamorphic Rock — Rock formed from pre-existing rocks that have been altered through an increase in temperature and/or pressure.

Migmatite — Thin layer, vein, or lens of coarse-grained granitic-type rock found within intensely metamorphosed rock.

Mineral — A homogeneous, naturally occurring, crystalline substance.

Monocline — Fold occurring in horizontally bedded rock that bends in one direction only, resulting in identical beds being found at different elevations on either side of the monocline.

Original Horizontality — Principle stating that sedimentary rocks are originally deposited horizontally.

Orogen — Zone of deformed metamorphic rock resulting from mountain building processes.

Permeability — The ability of a formation to transmit groundwater.

Plateau — An elevated area of relatively flat land that contains somewhat abrupt descents along much of its perimeter.

Playa — Lake bed, found in arid area, which is dry throughout most of the year.

Plutonic Rock — Igneous rock formed below the earth's surface at depth.

Regression — A continual lowering of sea level over a long interval of geologic time.

Rock — Naturally formed aggregate of one or more types of minerals.

Sandstone — Sedimentary rock comprised of cemented sand-sized fragments (mostly quartz).

Schist — A coarse-grained foliated metamorphic rock.

Sedimentary Rock — Rock comprised of weathered and eroded remnants of pre-existing rock that have been redeposited, in a horizontal manner.

Shale — Sedimentary rock consisting of silt and clay-sized fragments.

Sill — A flat igneous intrusion that has intruded parallel to the bedding of the host rock.

Glossary

Springs — A location where groundwater flows onto the ground surface.

Strata — A layer of bedded rock.

Stratified — Layered.

Stratigraphy — The study of stratified rocks.

Stromatolites — Fossils of primitive algae representing the earliest colonial organisms that evolved in ancient seas.

Superposition — Principle stating that rocks on the bottom of a bedded sequence are the oldest and become progressively younger towards the top of the sequence.

Trace Fossils — Indirect fossil evidence, such as tracks, trails, and burrows.

Transgression — A continual rise in sea level over a long interval of geologic time.

Travertine — A type of limestone deposited in fresh (as opposed to salt) water.

Trilobites — Invertebrate animals that lived on the sea floor in the early Paleozoic and were covered with a protective shell, much like a horseshoe crab.

Unconformity — A substantial gap in geologic time generally identified by an ancient erosional surface.

Volcanic Rock — Igneous rock formed on the earth's surface through volcanic eruption.

REFERENCES

Beus, Stanley and Michael Morales; 1990; *Grand Canyon Geology*; Oxford University Press.

Chronic, Halka; 1997; *Pages of Stone – Grand Canyon and the Plateau Country*; The Mountaineers.

Collier, Michael; 1980; *An Introduction to Grand Canyon Geology*; Grand Canyon Natural History Association.

Grand Canyon Association; 1996; *Geologic Map of the Eastern Portion of the Grand Canyon National Park*; Grand Canyon Association.

Harris, Ann, Esther Tuttle, and Sherwood Tuttle; 1990; *Geology of National Parks*; Kendall/Hunt Publishing Co.

Johnson, P. and R. Sanderson; 1968; *Spring Flow into the Colorado River, Lees Ferry to Lake Mead, Arizona*; Arizona State Land Department Water-Resources Report No. 34.

Lucchitta, Ivo; 1988; *Canyon Maker – A Geological History of the Colorado River*; Museum of Northern Arizona.

Shimer, John; 1972; *Field Guide to Landforms in the United States*; The Macmillan Company.

References

Thayer, Dave; 1986; *A Guide to Grand Canyon Geology Along Bright Angel Trail*; Grand Canyon Natural History Association.

Wicander, Reed and James Monroe; 1993; *Historical Geology*; West Publishing Company.